Vinícius Rodrigues
Daniel Mantovani
Aline T. A. Baptista

Preliminary risk analysis applied in a textile industry

Vinícius Rodrigues
Daniel Mantovani
Aline T. A. Baptista

Preliminary risk analysis applied in a textile industry

Application of a case study

Imprint

Any brand names and product names mentioned in this book are subject to trademark, brand or patent protection and are trademarks or registered trademarks of their respective holders. The use of brand names, product names, common names, trade names, product descriptions etc. even without a particular marking in this work is in no way to be construed to mean that such names may be regarded as unrestricted in respect of trademark and brand protection legislation and could thus be used by anyone.

Cover image: www.ingimage.com

This book is a translation from the original published under ISBN 978-613-9-65593-9.

Publisher:
Sciencia Scripts
is a trademark of
Dodo Books Indian Ocean Ltd. and OmniScriptum S.R.L publishing group

120 High Road, East Finchley, London, N2 9ED, United Kingdom
Str. Armeneasca 28/1, office 1, Chisinau MD-2012, Republic of Moldova, Europe
Printed at: see last page
ISBN: 978-620-7-79149-1

SUMMARY

SUMMARY

The country's clothing industry emerged from the process of national industrialization with textile production. As the textile industry evolved, the clothing sector modernized, and today there are twenty-one different segments of clothing and accessories. Occupational safety can be understood as a set of sciences and technologies that aim to protect workers in their working environment, seeking to minimize and/or avoid accidents at work and physical and mental occupational illnesses. The Preliminary Risk Analysis (PRA) is a study carried out during the planning phase of a project or system, with the aim of identifying the probable risks that may arise during its implementation and operation phase. Although it is usually carried out during the initial phase of a company's project, it is very effective for assessing risks that are already underway. This work is a case study of a textile industry that receives raw fabric and dyes it to prepare it for the garment industry, with the implementation of APR to assess the ongoing risks in the company. Thus, each stage of the company's process was analyzed, and the risks existing in each of the sectors studied were identified. Once the risks were identified, their severity was assessed and an action plan was drawn up to eliminate and/or reduce them as much as possible in order to improve the working environment and employee safety.

Key words: occupational safety; preliminary risk analysis; textile industry.

1 INTRODUCTION

The textile industry, along with mining and steel, formed the pillars of the industrial revolution in the 18th century. It was with the strategy of mechanizing processes in order to reduce costs that the system of capitalism in production began. The first country to industrialize in this way was England, a pioneer in the industrial revolution. In Brazil, this process began later, with the planting of cotton in the state of Maranhao, focused especially on exports to European countries, mainly to England itself. The first commercial relationships in the textile industry in Brazil began with the arrival of European immigrants, who already had baggage and knowledge of the industry, and began to found small family businesses in Brazil, which have developed into large textile companies (REZENDE FILHO, 2005; FURTADO, 2007).

According to Votorantim Metais (2005), Occupational Safety can be understood as the science that studies the possible causes of accidents, using techniques and methodologies, with the aim of preventing the occurrence of work-related accidents and illnesses, thus helping the quality and physical integrity of the employee, as well as advising the employer, preventing costs and problems due to possible accidents. In this sense, an accident at work is understood to be anything that occurs as a result of carrying out this work in the company's service. Safety aims to eliminate and/or reduce these events to a minimum, avoiding short- and long-term bodily injuries that could damage the employee's quality of life and performance. It can also be said that accidents are unplanned events that have the power to affect a worker's performance or waste material (VOTORANTIM METAIS, 2005).

Occupational Safety and Health (OSH), which establishes a legal and social link with the well-being of employees, as well as preventing work-related accidents, injuries and illnesses, helps to control and treat them.

For Souza (2011), the improvement that has been observed is due to the constant reform and revision of the legal requirements related to work safety, which is not yet effectively reflected in the field, given the high accident rates that are recorded in companies, This demonstrates the need for a more present and active attitude on the part of those responsible for the prior assessment of the risks to which employees are exposed, risks which lead companies to suffer stoppages, reductions and/or freezing of profits, even the death of employees, if proper attention is not paid to this issue.

Tavares (2004) defines APR as a preliminary study, during the planning phase of a process, where the main objective is to identify the risks present, their source, and with this data carry out planning and/or management to control the risks.

Among the existing ways and methods for identifying the risks to which workers are exposed in the workplace is the APR, which, together with good risk management, are extremely important tools for reducing and controlling risks.

2 OBJECTIVES

Carrying out a Preliminary Risk Analysis (PRA) in a small/medium-sized textile dyeing industry.

2.1 SPECIFIC OBJECTIVES

The specific objectives will be focused on identifying the processes and activities that exist in the fabric dyeing process along the production flowchart:

-	Checking the possible risks present in the production plant;

-	Analyze and establish the hierarchy of risks;

-	Suggest an action plan to control the risks.

3 BACKGROUND

Nowadays, OSH is a recurring issue in companies around the world, but in Brazil it is still a long way from being more widely disseminated. In 2019, e-Social will come into force, a program in which a system will be set up so that employers will be able to communicate to the government, in a unified way, information about their workers, such as employment relationships, social security contributions, payroll, communications about accidents at work, prior notice, tax deeds and information about the Severance Indemnity Fund (FGTS). As a result, companies are paying more attention to the OSH issue.

3.1 PRELIMINARY RISK ANALYSIS

The APR is an analysis carried out in the planning process of a production system, a company, with the aim of identifying the possible risks that would be present during its implementation and execution. The system is effective in the planning phase, however, its use is also very important in processes that are already underway, providing an analysis of the quality of the working environment, showing risks that have not previously been analyzed.

According to Zocchio (2000), APR began in the military, where preliminary reviews were carried out on missile launches. Its main objective is to determine what risks are present in the work environment, and to adopt measures that can prevent and/or extinguish risks, even before the process is implemented, i.e. to apply an analysis during the process development phase. Every stage must be analyzed, and every form of risk to the physical and mental integrity of the employee must be taken into account, so that, through risk management policies and plans, adversities can be avoided.

The APR consists of identifying and analyzing all the risks that exist in the workplace, their causes and consequences for the worker, and establishing methods that can control the danger. One of its characteristics is that it requires a multidisciplinary team, i.e. extensive knowledge in various areas, so that all the risks can be identified, as well as in-depth knowledge of all the processes involved in the work environment, such as the employees themselves.

Amorim (2013) points out that the main advantages of RPA are the early identification of risks and raising awareness of the potential dangers in each process among the entire team, so that they can work on projects to combat them. In this way, as the projects develop, the main dangers encountered are eliminated and/or reduced even before the company's processes begin to be implemented.

3.2 RISK MANAGEMENT

Gonçalves (2000) succinctly defines risk management as the application of a system of strategies, which comes from a dynamic project that brings together the whole process from identifying risks, analyzing them and drawing up a control plan. Its main objective is to reduce the risks that are a consequence of the process itself, or by means of equipment. As a result, this risk control policy is able to meet the requirements imposed by occupational safety, since it can control risks and identify faults in the working environment.

Thus, the set of actions involving RPA and risk management makes it possible to prevent accidents, injuries and work-related illnesses and, as a result, to reduce companies' costs and expenses. The application of these procedures reflects on the overall organization of the company as a whole, as it affects internal and external means, i.e. within the company's procedures and layout, as well as financial factors, political and economic agreements.

Therefore, relating concepts makes companies aware of and plan an effective risk management system, designed to prevent all risks, using interpersonal relationships to build a safe and healthy working environment (LAPA and GOES, 2011).

Based on NBR ISO 31010 (2012), risk management policies make it possible to reduce the costs involved in accidents and occupational illnesses by preventing accidents in the workplace.

4   MATERIALS AND METHODS

4.1 DESCRIPTION OF THE COMPANY

The company studied operates in the textile sector, where it dyes and prepares raw fabric for the clothing industry. Despite having a large physical area, the company has a relatively small workforce, with approximately 8 to 10 employees per shift, since the company operates 24 hours a day.

Figure 01 shows a sketch of the company's internal facilities, except for the boiler, which will be discussed later:

Figure 1 - Sketch of the company area.

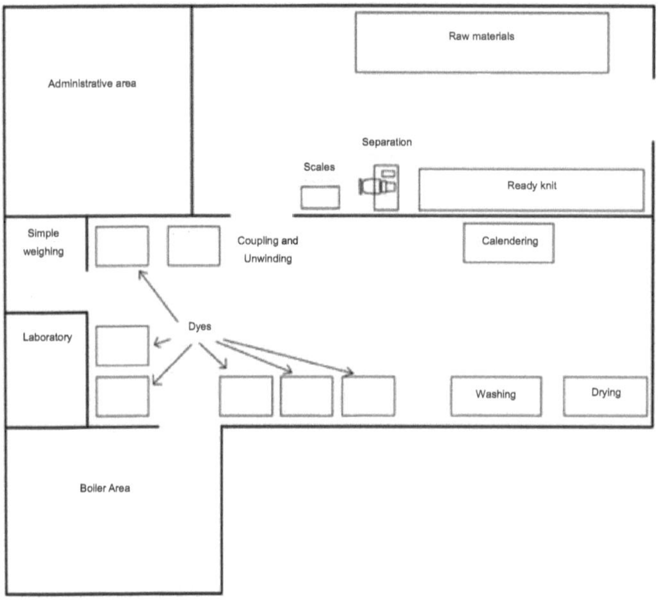

The company's most striking feature, apart from its integral operation, is the boiler feed (pressure vessel).

The process carried out in the company is practically all mechanized, with little human contact, only for small parts of the process, such as transporting the material between machines, weighing the meshes and dyes, among others.

4.1 CONCEPTS

Chemical agents: According to NR 9, these are "substances, compounds or products that can enter the body via the respiratory route, in the form of dust, fumes, mists, fogs, gases or vapors, or which, due to the nature of the exposure activity, can come into contact with or be absorbed by the body through the skin or by ingestion".

Physical agents: According to NR 9, these are "the various forms of energy to which workers may be exposed, such as: noise, vibrations, abnormal pressures, extreme temperatures, ionizing radiation, non-ionizing radiation, as well as infrasound and ultrasound".

Risk analysis: According to ISO *Guide* 73 (2009), this is the process of understanding the nature of the risk and determining the level of risk.

Preliminary risk analysis: According to NR 33, this is the "initial assessment of potential risks, their causes, consequences and control measures".

Unsafe act: According to NBR 14280, it is "an action or omission which, contrary to safety requirements, may cause or favor the occurrence of an accident".

Risk identification: According to ISO *Guide* 73 (2009), this is the process of finding, recognizing and describing risks.

Risk source: According to ISO *Guide* 73 (2009), it is the element that alone or in combination has the intrinsic potential to give rise to a risk.

Risk management: According to ISO *Guide* 73 (2009), it is the coordinated activities to direct and control an organization in relation to risks.

Risk analysis methodologies: According to NR 20, these are "methods and techniques which, when applied to operations involving process or processing, identify hypothetical scenarios of undesired occurrences (accidents), the possibilities of damage, effects and consequences".

Examples of some methodologies:

a) Preliminary Hazard/Risk Analysis (APP/APR);

b) "What-if";

c) Risk and Operability Analysis (HAZOP);

d) Failure Modes and Effects Analysis (FMEA/FMECA);

e) Fault Tree Analysis (FCA);

f) Event Tree Analysis (EFA);

g) Quantitative Risk Analysis (QRA).

Hazard: According to NBR 14726, it is the "situation with the potential to cause personal injury or damage to health, the environment or property, or a combination of these". Risk: According to NBR 14726, it is the "property of a hazard to promote damage, with the possibility of human, environmental, material and/or economic losses, resulting from the combination of expected frequency and consequence of these losses".

4.2 METHODS

The work was directed at a case study of a company in area x located in the northwestern region of Paranà, where each part of its operation was carried out by recording photos and describing sectors.

Each stage of the fabric dyeing process was individually monitored from start to finish, with a methodology applied to the procedures and risks encountered. Once this material had been collected, all the risks to which employees are exposed at each stage of the dyeing process, as well as when transporting the fabrics, were assessed and identified.

Once the risks had been identified, an analysis of the severity of each risk was carried out in order to draw up an action plan.

At the end of the procedure, the sources of risk were addressed, through the correct use of the Collective Protection Equipment (CPE) and Personal Protection Equipment (PPE) necessary for the safety of employees throughout the process.

5 DEVELOPMENT

5.1 PROCESSES AND RISK ASSESSMENT

The process carried out within the company, through all the work logistics, is programmed in the following format:

1. Receiving raw mesh (in rolls) and storage;

2. Sorting and weighing meshes for painting;

3. Joining and unjoining the meshes;

4. Preparing the dye recipe;

5. Mesh dyeing;

6. Washing the fabric;

7. Drying the mesh;

8. Calendering and preparation for shipment;

9. Transportation.

All the process-oriented contexts are shown in Figures 2 to 19, illustrated with the positive and negative points of the process:

5.1.1 - RECEIVING RAW MATERIALS

Figure 2 - Raw material (raw mesh).

Figure 3 - Unloading raw materials.

The company's first process is to receive the raw mesh in rolls, which arrive by truck. Of all the company's processes, this is one of the most problematic. The unloading of all the material is done manually by two employees, who unload the truck and stack the rolls of mesh on pallets, which will help with transportation in the next stages.

Although the weight of each roll is not that heavy, approximately 20kg, the way in which they are unloaded is very harmful and dangerous, especially at the beginning of the process, where one of the employees climbs on top of the pile of rolls inside the truck, which exceeds 2m and can already be considered work at height, and throws the roll to another employee, as can be seen in Figure 2. It was also noted that none of these employees wore any kind of PPE when carrying out the process.

Chart 1 - Pros and cons of receiving raw materials.

PROS	CONS
• The weight of the material is not excessive; • Storing the meshes on pallets to facilitate subsequent transportation.	• Lack of PPE and CPE; • Lack of training.

Table 2 - Raw material risks ERGONOMIC RISKS

- Tiring and/or excessive work;

- Lack of orientation and training;

- Repetitive movements;

- Inadequate and non-ergonomic equipment.

RISK OF ACCIDENTS

- Inadequate, defective or non-existent equipment;

- Risk of falling from a level, injuries due to impact with objects;

- Cargo and transportation in general.

5.1.2 - WEIGHING AND SEPARATING MESHES

Figure 04 - Sorting and weighing mesh for dyeing.

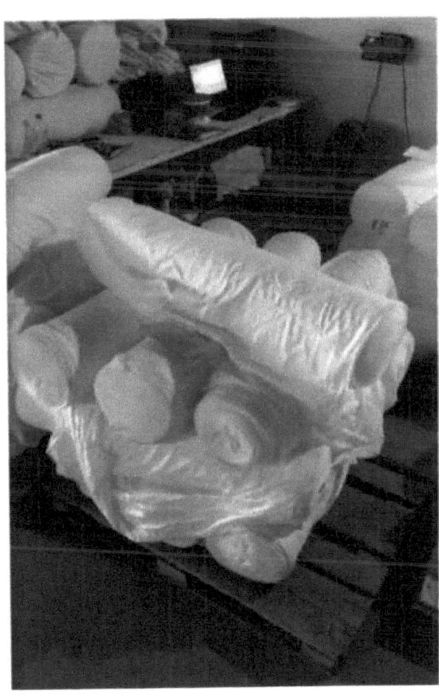

At this stage, the process is carried out by an employee who is responsible for receiving orders and work orders containing the type of knitwear, quantity and color of the knitwear. With this data, he is responsible for fetching the pallets with the raw fabric and weighing the fabric and separating it for the dyeing machines, which have different weight capacities. The pallets in this process, as in the others from this stage onwards, are transported by manual trolleys. Some points noted at this stage were that no PPE was used and the work chair did not have any ergonomic adjustments.

Table 3 - Pros and cons of separating and weighing meshes.

PROS	CONS
• Transporting the meshes by trolley; • Scale suitable for weighing.	• Lack of PPE and CPE; • Work chair with no adjustments; • The atmosphere is stuffy and hot at certain times.

Table 04 - Risks of separating and weighing meshes.

ERGONOMIC RISKS

• Lack of orientation and training;   • Repetitive movements;   • Inadequate and non-ergonomic equipment.   **RISK OF ACCIDENTS**   • Risk of falling from a level, injuries due to impact with objects;   • Cargo and transportation in general.

5.1.3 - JOINING AND UNKNOTTING

Figure 5 - Unwinding and joining the meshes.

Figure 6 - Unwinding and joining the meshes.

This process is carried out by just one employee like the previous one. This stage consists of receiving the material separated in the previous stage and then unwinding the mesh roll and joining the mesh between one roll and another to continue the mesh. From this stage onwards, when it comes to meshing procedures, the process is almost entirely automated. Here the employee loads the machine with a roll of mesh, which will be unwound and placed on top of the next pallet. At the end of a roll, the employee joins the mesh at the end of the first roll with the beginning of the next roll on a sewing machine, thus ensuring the continuity of the mesh. It was observed, as in the previous processes, that the employee did not wear PPE, although the stage does not present many risks, the use of PPE is essential.

Table 5 - Pros and Cons of unrolling and joining meshes.

PROS	CONS
• Very automated work; • Machine control panel compliant with standards.	• Lack of PPE and CPE; • Work chair with no adjustments; • The atmosphere is stuffy and hot at certain times;

	• No equipment manual available.

Table 06 - Risks of unwinding and joining meshes.

ERGONOMIC RISKS

- Repetitive movements;

- Inadequate and non-ergonomic equipment.

RISK OF ACCIDENTS

- Risk of falling from a level, injuries due to impact with objects;

- Cargo and transportation in general.

PHYSICAL RISKS

- Excessive cold and/or heat.

5.1.4 - RECIPE FORMULATION AND PREPARATION OF COLORANTS

Figure 7 - Preparing the dye recipe (on a scale) for dyeing.

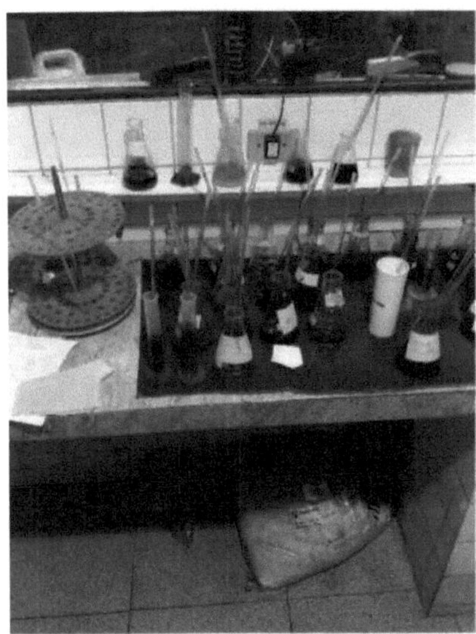

Figure 8 - Heavy dyes ready for mixing.

Figure 9 - Dye mixer.

Figure 10 - Dyeing machine control panel.

This stage takes place in a small, air-conditioned and ventilated laboratory. Here, the dye recipe is formulated on a scale, i.e. the amount of each dye needed to produce the color desired by the customer is calculated. Everything here is done on a scale for 10g of mesh, and once these calculations have been made, they are proportionally recalculated for the weight of the mesh to be dyed on each machine. The final weighing process of the dyes for dyeing is done in a room next door, where the dyes are stored. Once the dyes have been weighed, they are placed in the dyeing machine, where the recipe for the desired color is entered electronically and the dyeing process begins.

Unlike the previous stages, in this process it was observed that the employees involved wore PPE, both in the laboratory and in the weighing room, and as these are chemical materials, the control of this process appeared to be more controlled.

Table 7 - Pros and cons of formulating the recipe and preparing the dye.

PROS	CONS

• Use of PPE;	• Open weighing room (with no effective restrictions on people entering);
• Restricted access environment (laboratory);	• Unsuitable containers for the dyes.
• Air-conditioned environment;	
• Adequate signage;	
• Suitable machinery, with control panel in accordance with the standard;	
• Well-automated process.	

Table 8 - Risks of formulating the recipe and preparing the dye.

ERGONOMIC RISKS

- Lack of orientation and training;

- Repetitive movements;

- Inadequate and non-ergonomic equipment.

RISK OF ACCIDENTS

- Risk of falling from a level, injuries due to impact with objects;

- Cargo and transportation in general.

CHEMICAL RISKS

- Solvents (especially volatile ones);

- Acids, bases, salts, alcohols, ethers, etc;

- Chemical reactions in general.

5.1.5 - FABRIC DYEING

Figure 11 - Dyeing machine.

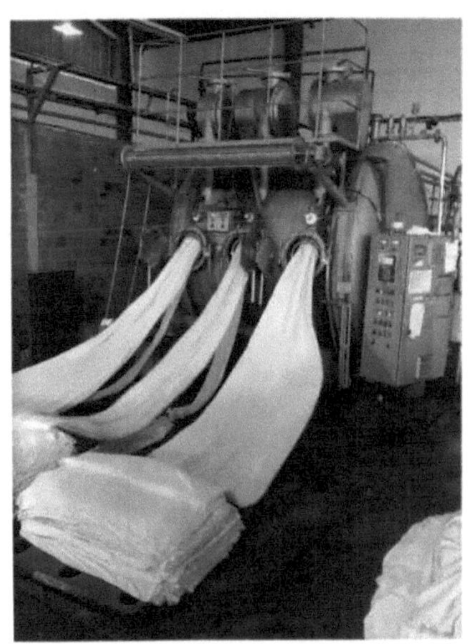

Figure 12 - Dyed knitwear ready for washing.

With the dyes weighed and prepared, the knitwear that was previously unrolled and joined together is brought to the dyeing machines, which, once fed, carry out the whole process on their own, delivering the knitwear completely dyed and ready for the next stage. It was observed that, as in other stages, here the employee does not wear PPE.

Chart 09 - Pros and cons of dyeing knitwear.

PROS	CONS
• Very automated work; • Machine control panel compliant with standards.	• Lack of PPE and CPE; • The atmosphere is stuffy and hot at certain times; • No equipment manual available.

Chart 10 - Risks of dyeing knitwear.

PHYSICAL RISKS

• Excessive cold and/or heat.

RISK OF ACCIDENTS
• Risk of falling from a level, injuries due to impact with objects; • Cargo and transportation in general.

5.1.6 - MESH WASHING

Figure 13 - Washing machine.

Figure 14 - Washed mesh, ready for the drying process.

Once the fabric has been dyed, it is taken to the washing process. In this process, the fabric is washed to remove the access of dye and impurities contained in the fabric, in addition to the addition of some products to prepare the fabric, such as softeners, so that the fabric begins to gain a softer texture that is observed in the finished fabric. At the end of the wash, the fabric is placed on another pallet, folded, for the next procedure.

At this stage, it was observed that the employee was using PPE correctly, since he was constantly in contact with water, as well as some chemicals, due to the products inserted in the machine.

Chart 11 - Pros and cons of mesh washing.

PROS	CONS
• Very automated work; • Machine control panel compliant with standards.	- No equipment manual available.

Table 12 - Mesh washing risks.

PHYSICAL RISKS
• Excessive cold and/or heat;

• Humidity. RISK OF ACCIDENTS • Risk of falling from a level, injuries due to impact with objects; • Cargo and transportation in general. **CHEMICAL RISKS** • Acids, bases, salts, alcohol, ether, among other chemicals; • Chemical reactions in general.

5.1.7 - DRYING THE MESH

Figure 15 - Drying process.

Figure 16 - Dry mesh, ready for the calendering process.

Once the mesh has been washed, the pallet is transported to the next machine, where the mesh will be dried. This is a very simple process, where the machine is fed, then the mesh passes along a conveyor belt through the machine, and at the end of the drying process, it is deposited on a pallet, folded, for the next and final procedure on the mesh. As with other processes, the employee in this process didn't even wear PPE.

Chart 13 - Pros and cons of mesh drying.

PROS	CONS
• Very automated work; • Machine control panel compliant with standards.	• Lack of PPE and CPE; • The atmosphere is stuffy and hot at certain times; • Some moving parts exposed; • No equipment manual available.

Table 14 - Mesh drying risks.

PHYSICAL RISKS
• Excessive cold and/or heat.

RISK OF ACCIDENTS
• Risk of falling from a level, injuries due to impact with objects; • Cargo and transportation in general; • Risk of burns.

5.1.8 - CALENDERING AND PREPARATION FOR SHIPMENT

Figure 17 - Calendering process.

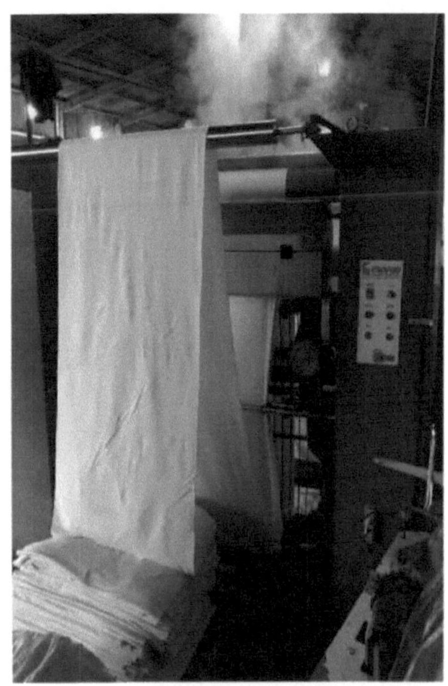

Figure 18 - Finished mesh being rolled up.

The process of dyeing the mesh ends with its passage through the "Calender", where the dried material is passed through, cut, and rolled up again for shipment. As with most stages, it consists of feeding the machine, which passes and cuts the mesh to the size desired by the customer, and then rolls it up into 20 kg rolls. At the end of this stage, the employee inserts the roll into a plastic bag to protect the mesh, and stacks it on a pallet so that it can be sent for transportation. As with other stages, the use of PPE was not observed here.

Chart 15 - Pros and cons of calendering and preparation for shipment.

PROS	CONS
• Very automated work; • Machine control panel compliant with standards.	• Lack of PPE and CPE; • The atmosphere is stuffy and hot at certain times; • No equipment manual available.

Chart 16 - Calendering risks and preparation for shipment.

PHYSICAL RISKS

•   Excessive cold and/or heat.

RISK OF ACCIDENTS
•   Risk of falling from a level, injuries due to impact with objects;
•   Cargo and transportation in general;
•   Risk of burns.

5.1.9 - PRESSURE VESSEL BOILER

To power the entire process, the company uses a boiler, a combustion pressure vessel, located outside the company, next to the machine line, as shown in Figure 1.

A visit was made to the boiler, and some important points were noted about it, since it is the part of the whole company that is most at risk and needs the most attention. The images below show what the company's boiler and its feed looks like:

Figure 20 - Boiler.

Figure 21 - Boiler from the side.

Figure 23 - Wood for feeding the boiler.

To assess the boiler, a *checklist* was carried out *in* accordance with NR-13, which deals with pressure vessels:

Chart 17 - Boiler checklist.

PRESSURE VESSEL CHECKLIST ACCORDING TO NR-13	YES	NO
1- Is the equipment working below the Maximum Allowable Working Pressure (MAWP)?	J	
2-It constitutes a serious and imminent risk if any of the following items are missing:		
a) Is the safety valve in good condition?	J	
b) Is the Safety Valve set to an opening pressure equal to or less than PMTA?	J	
c) Is the pressure gauge in good working order?	J	

Question			
d) Is there a fast and well-functioning drainage system?	J		
e) Indication system to control the water level and prevent overheating due to poor supply, working well?	J		
3-Boiler body			
a) Is the boiler nameplate in a visible and easily accessible place?	J		
b) Does the nameplate on the boiler contain information such as the manufacturer's name, year of manufacture, PMTA, hydrostatic test pressure, steam production capacity and boiler category?	J		
4-Are there two emergency exits, wide and in designated directions?	J		
5-Is the ventilation permanent with air inlets that cannot be blocked?	S		
6-Is there easy and safe access needed to operate and maintain the boiler?	J		
7-Is there a system for capturing and releasing gases and particulate matter from combustion outside the operating area?	J		
8-Is there lighting in accordance with current regulations and emergency lighting?		J	
9-Is there a boiler operating manual?	J		
10-Are you continuously at your workstation?	J		
11-Do boiler operators have a boiler safety training certificate?		J	
12-Practical training ?		J	
13-Is there an inspection report issued by a qualified professional?	J		
14-Are there inspection records issued by a qualified professional?	J		

15- Is there fire protection according to NR-23?	J	
16-Does it comply with NR-20, combustible and flammable liquids?		J

Table 18 - Pros and cons of the boiler.

PROS	CONS
• Regulated boiler • Machine control panel compliant with standards	• Lack of PPE and EPCs • Hot and stuffy environment • Disorganized wood for boiler feed • Boiler feed • Inappropriate layout

Table 19 - Boiler risks.

ERGONOMIC RISKS

- Tiring and/or excessive work

- Poor body posture in relation to work

PHYSICAL RISKS

- Excessive cold and/or heat

RISK OF ACCIDENTS

- Risk of falling from a level, injuries due to impact with objects

- Cargo and transport in general

- Risk of burns, fire, explosion

BIOLOGICAL RISK

- Risk of insects in firewood

- Risk of microorganisms (viruses, bacteria, protozoa) due to firewood, and possible nails in the wood

5.1.10 - SIGNS AND SAFETY EQUIPMENT

In an interview with the employee, it was found that the company provided individual safety equipment when each employee was hired:

- Goggles;

- Rubber glove;

- Rubber boots;

- Earplugs.

Although it has been provided, there is no control over the use of this PPE by the foremen, and there is no control over the expiration date or condition of the equipment. The interview also revealed that there are no courses on how to use this equipment correctly, i.e. although PPE is provided, the lack of instruction on how to use it invalidates its presence.

In terms of EPC, it was noted that the company has several signposts throughout its layout, as well as restricted access in some sectors, such as the chemical laboratory and the boiler room. The power panel is well signposted and there is no exposed wiring. Some of these signs can be seen in the images below:

Figure 25 - PPE sign.

Figure 26 - Boiler sign.

Figure 27 - Electrical panel and signpost.

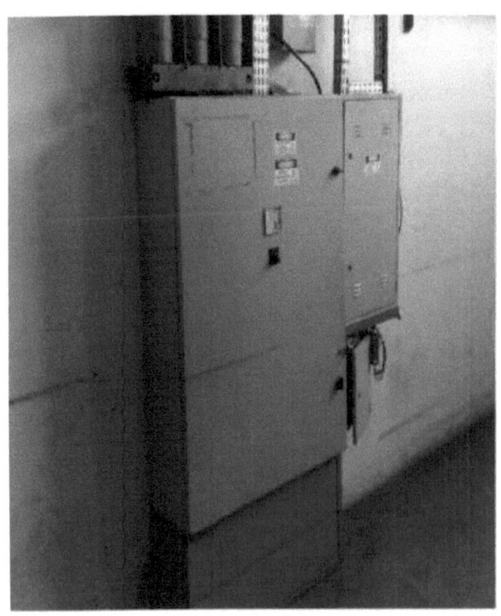

5.2 - ACTION PLAN

Based on the visit and the data collected, it was possible to identify a number of flaws and risks in the company's line of work. The seriousness of these risks for employees was then assessed, and based on this seriousness, an Action Plan was drawn up, which the company must follow in order to eliminate the risks presented.

The Action Plan, in order of priority, can be seen in Table 20.

Table 20 - Action Plan

Training of employees responsible for the boiler.

Improved organization and layout of the boiler.

Acquisition of suitable PPE for boiler workers, such as:

- Cowhide gloves;

- Cowhide aprons;

- Safety glasses;

- Safety boots.

Purchase of PPE for employees responsible for receiving raw materials, such as:

- Helmet;

- Protective gloves and/or cream;

- Protective boots;

- Lumbar protector.

Training on the importance and use of PPE for all employees.

Guiding and supervising the use of PPE by employees.

Restrict access to the dye and chemical weighing room to authorized personnel only.

Purchase of chairs with ergonomic adjustments.

Improve air conditioning.

6 CONCLUSION

The proximity to the proposed case study, as well as the analysis carried out in the textile company, addresses the importance of OSH in the most diverse Brazilian companies and their many segments. Thus, during the study it was possible to see that although the company has up-to-date equipment, where the risk to the employee is extremely low, provides PPE for its employees and has an EPC policy, it does not have any kind of training or preparation for its employees, which means that the risk of accidents is high. The company operates 24 hours a day, in three 8-hour shifts, and the training of all its employees is necessary.

The area that presented the greatest risk was the boiler, which, being a pressure vessel, is one of the most dangerous devices in companies. Although the conditions for quality and certification of the boiler were in line with the prescriptions of the Regulatory Standard, none of the operators had the knowledge attributed to the necessary training described in the standard, to work in this sector, and the layout of the boiler is not very favorable, especially in the arrangement of the ignition material, wood and firewood to feed the furnace with minimal organization. In order to reduce problems of inefficiency and risks to workers' occupational health, the action plan was drawn up to eliminate or reduce as far as possible the risks and problems encountered, thus providing safer and more comfortable working conditions for employees and the company as a whole.

7 REFERENCES

AGUIAR, L. A. **Risk Analysis Methodologies APP & HAZOP**. Rio de Janeiro, UFRJ, 2011.

AMORIM. E.L.C. **Risk Analysis Tools Workbook.** Maceió: UFAL, 2013.

DE CICCO, F.; FANTAZZINI, M. L. **Established risk management technologies**. 2 ed. Sao Paulo: Risk Tecnologia, 2003.

FARIA, M. T. **Gerência de Riscos: Apostila do curso de Especialização em Engenharia de Segurança do Trabalho**. Curitiba, Paranâ. UTFPR, 2011.

FRUHAUF, D. V.; CAMPOS, D. T. A.; HUPPES, M. N. **Aplicaçao da ferramenta anàlise preliminar de riscos** - estudo de caso indùria frigorifica de frangos. 2005. 42f. Monograph for the Postgraduate Course in Occupational Safety Engineering at the State University of Ponta Grossa, 2005.

FUNDACENTRO, Fundaçao Jorge Duprat Figueiredo de Segurança e Medicina do Trabalho. **Introduction to Occupational Hygiene.** Sao Paulo: FUNDACENTRO, 2004.

FURTADO, Celso. **Economic Formation of Brazil.** 34. ed. Sao Paulo: Companhia das Letras, 2007.

GOMES, P. C. dos R.; OLIVEIRA, P. R. A. de. **Introduçao à Engenharia** de **Segurança do Trabalho.** Brasilia: WEducacional e Cursos LTDA, 2012.

GONÇALVEZ, Edwar Abreu. **Occupational Safety and Medicine in 1,200 Questions and Answers**. 3ª Ed. Sao Paulo, Editora LTR, 2000

LAPA, R. P.; GOES, M. L. S.; **Investigation and analysis of incidents**. Knowing the incident to prevent it, 1 ed. Sao Paulo, EDICON, 2011.

MTE. Ministry of Labor and Employment. **Atlas Manuals of Legislation.**

Occupational Safety and Medicine. Sao Paulo: Atlas, 2017.\

OLIVEIRA, C.A.D. **Manual Pràtico de Saù e Segurança do Trabalho**. Sao Caetano do Sul, Sao Paulo: Yendis Editora, 2010.

REZENDE, F., Barros, C.. **General Economic History.** 8 Ed. Editora Contexto, Sao Paulo, SP, 2005.

VOTORANTIM METAIS. Votorantim **management system.** Observer's Manual. 1 .ed. Juiz de Fora, MG, 2005.

SOUZA, C.R.C. **Analyzing and Managing Risks in Industrial Processes**. Risk Management Workbook. Niterói, Rio de Janeiro, UFF, 2011

TAVARES, José da Cunha. **Noçoes de Prevençao e controle de perdas em segurança do trabalho**. Sao Paulo: Senac, 2004.

ZOCCHIO, A. Política de segurança e saùde no trabalho. Editora LTR, Sao Paulo: SP, 2000.

Printed by Books on Demand GmbH, Norderstedt / Germany